DINOSAUR FOSSILS

BY NATALIE HUMPHREY

Published in 2025 by The Rosen Publishing Group
in association with Scientific American Educational Publishing
2544 Clinton Street, Buffalo NY 14224

Library of Congress Cataloging-in-Publication Data
Names: Humphrey, Natalie author
Title: Dinosaur fossils / Natalie Humphrey.
Description: Buffalo, NY : Enslow Publishing, [2025] | Series: Scientific
   American investigates fossils | Includes index. | Audience term:
   Juvenile
Identifiers: LCCN 2024038056 (print) | LCCN 2024038057 (ebook) | ISBN
   9781725351967 (library binding) | ISBN 9781725351950 (paperback) | ISBN
   9781725351974 (ebook)
Subjects: LCSH: Dinosaurs–Juvenile literature
   http://id.loc.gov/authorities/subjects/sh2008102205 | Reptiles,
   Fossil–Juvenile literature
   http://id.loc.gov/authorities/subjects/sh2010110818 | Fossils–Juvenile
   literature http://id.loc.gov/authorities/subjects/sh2008104067
Classification: LCC QE861.5 .H86 2025 (print) | LCC QE861.5 (ebook) | DDC
   567.9–dc23/eng/20240816
LC record available at https://lccn.loc.gov/2024038056
LC ebook record available at https://lccn.loc.gov/2024038057

Portions of this work were originally authored by Kathleen Connors and published as *Dinosaur Fossils*. All new material in this edition is authored by Natalie Humphrey.

Designer: Andrea Davison-Bartolotta
Editor: Natalie Humphrey

Some of the images in this book illustrate individuals who are models. The depictions do not imply actual situations or events.

Printed in the United States of America

CPSIA compliance information: Batch #CWSA25. For Further Information contact Rosen Publishing at 1-800-237-9932.

CONTENTS

Words in the glossary appear in **bold** type the first time they are used in the text.

HOW TO STUDY A DINOSAUR

Dinosaurs have been gone for over 65 million years, but we can still study them! Paleontologists are scientists who study what the Earth looked like long ago by looking at fossils.

By studying different kinds of fossils, paleontologists can put together a picture of what dinosaurs may have looked like. They can discover what dinosaurs may have eaten and how they may have lived. Some paleontologists have even found what dinosaurs may have sounded like.

FUN FACT

FOSSILS ARE THE MARKS OR REMAINS OF PLANTS AND ANIMALS THAT FORMED OVER THOUSANDS OR MILLIONS OF YEARS.

Dinosaur fossils are found everywhere on Earth. Most have been found in the deserts of North America, China, and Argentina.

SKELETAL FOSSILS

Over a long period of time, the living parts of a dinosaur's bones break down. The nonliving parts of the bones are left behind. Sometimes, **minerals** fill the tiny spaces in the bones. Other times, minerals take the place of the bones entirely. The result is a skeletal fossil!

Skeletal fossils include claws, teeth, and other hard parts of a dinosaur. Scientists have never found a fully intact, or complete, dinosaur skeleton!

Many museums
display dinosaur
skeletons. Often,
these are copies
of skeletal fossils.

FUN FACT

MOST DINOSAUR SKELETONS
ARE MISSING PIECES WHEN
THEY'RE FOUND.

TRACES OF DINOSAURS

Paleontologists also learn about dinosaurs through trace fossils. Trace fossils **preserve** facts about how dinosaurs behaved, or acted. **Burrows** and tooth marks are trace fossils. These are often found in rock made from **sediment**.

Some of the best-preserved trace fossils are dinosaur tracks. They can tell paleontologists a lot about dinosaurs. From the spacing of the tracks, paleontologists can tell if the dinosaur was walking or running. They can also tell if the dinosaur was moving with a group or moving alone.

FUN FACT

PALEONTOLOGISTS OFTEN CAN'T TELL THE TYPE OF DINOSAUR FROM ITS TRACKS. HOWEVER, THEY CAN TELL IF A DINOSAUR WAS MOVING ON TWO LEGS OR FOUR LEGS.

HUMAN FOSSILS OR DINOSAUR FOSSILS?

Paleontologists sometimes have a hard time telling what kinds of animals different fossils come from. Long ago, they thought many early dinosaur discoveries were actually human bones! In 1677, a scientist from England thought he'd found a bone from a giant human. He had found part of a dinosaur's thigh bone!

Years later, in 1818, a man named Samuel Ellsworth Jr. found some bones while digging a well in Connecticut. People thought the bones were human, but they were really from a dinosaur called *Anchisaurus*.

Early scientists believed dinosaurs looked much different than how scientists imagine them today.

FUN FACT
FOSSILS ARE OFTEN HEAVY, DARK IN COLOR, ROUGH TO THE TOUCH, AND HAVE MANY SMALL HOLES CALLED PORES.

DINOSAURIA

Many fossils and pieces of fossils were found in southern England throughout the 1800s. A scientist named Richard Owen thought the large bones came from **reptile**-like animals that were a similar species, or kind. He called them Dinosauria in the 1840s.

The fossil pieces didn't show what dinosaurs looked like. In 1834, a scientist found an incomplete skeleton near the English village of Maidstone. It was named the Maidstone *Iguanodon*. This skeleton gave scientists their first idea of what dinosaurs may have looked like.

FUN FACT

OWEN BELIEVED THERE WERE THREE GROUPS OF DINOSAURS: MEAT EATERS, PLANT EATERS, AND DINOSAURS WITH **ARMOR**. THESE GROUPS ARE STILL USED TODAY.

HOW OLD?

The oldest dinosaur fossils date back to around 240 million years ago. To find out how old a fossil is, scientists can use **superposition**. This establishes the fossil's age based on which **layer** of rock it's found in.

Another method is called radiometric dating. Scientists look at the elements usually found in fossils. Different elements break down at different rates, or speeds. By looking at the different breakdown rates, scientists can get a better idea of exactly how old a fossil is.

The youngest dinosaur fossils are from around 65 million years ago.

A DINOSAUR'S DINNER

Scientists can tell what a dinosaur ate from fossils of their teeth. Dinosaurs with sharp, pointed teeth commonly ate meat. Dinosaurs with flat, wide teeth usually ate plants. Some dinosaurs ate both! These dinosaurs had both flat and sharp teeth.

Footprint fossils can show what a dinosaur ate too! A footprint with three toes and sharp claws probably belonged to a meat-eating dinosaur. A footprint with three rounded toes probably belonged to a plant-eating dinosaur.

The dinosaur with the most teeth ever discovered, Nigersaurus, had over 500 teeth!
NIGERSAURUS

BREAKING NEWS!

Paleontologists are still finding new fossils and new ways to study them! In January 2011, Japanese scientist Tai Kubo reported he used footprints to find the weights of large, winged reptiles called pterodactyls (tehr-uh-DAK-tuhls). He studied the footprints of animals living today to help him.

In February 2024, scientists found the fossil of a large pterodactyl on the Isle of Skye in Scotland. It may have lived 166 to 168 million years ago!

Tai Kubo
found that
pterodactyls
weighed up to
320 pounds
(145 kg)!

VISIT A DINOSAUR

Would you like to see a dinosaur fossil in person? The American Museum of Natural History in New York City has many dinosaur fossils for you to see. You can also visit parks such as Dinosaur Valley State Park in Texas or Dinosaur National Monument in Colorado to see more dinosaur fossils.

You don't need to be a **professional** to find dinosaur bones either. Many everyday people have found dinosaur bones. You could make the next discovery!

Dinosaurs in National Parks

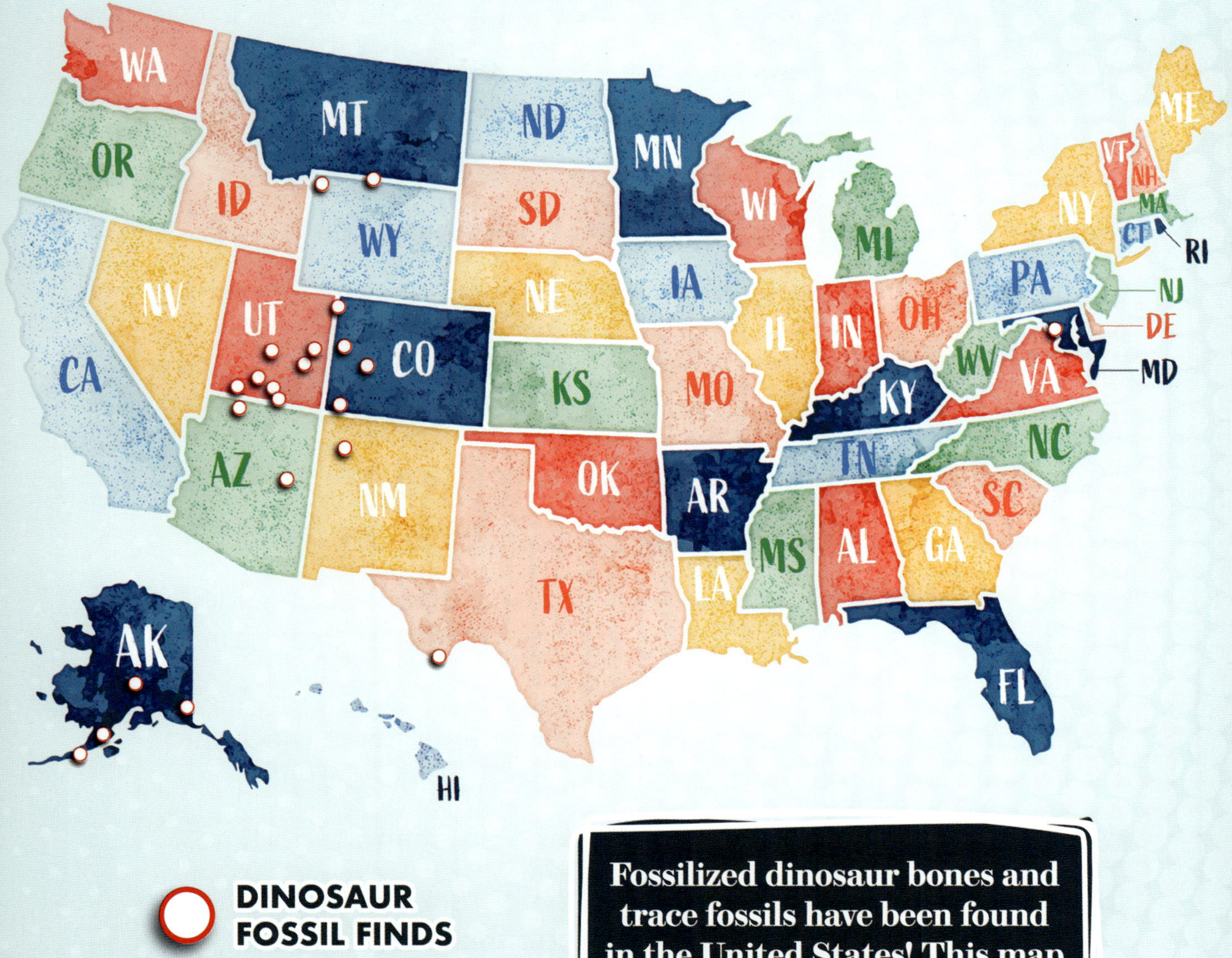

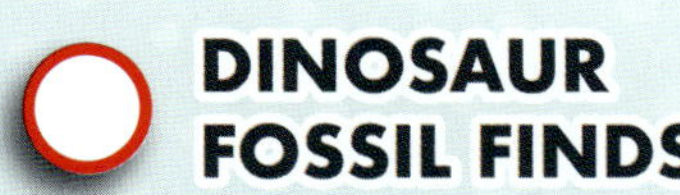

Fossilized dinosaur bones and trace fossils have been found in the United States! This map shows where fossils have been found in U.S. national parks.

GLOSSARY

armor: An animal's hard covering that protects its body.

burrow: A hole made by an animal in which it hides or lives.

layer: One thickness of something lying over or under another.

mineral: Matter in the ground that forms rocks.

museum: A building in which things of interest are displayed.

preserve: To keep something in its original state.

professional: Having to do with a job someone does for a living.

reptile: An animal covered with scales or plates that breathes air, has a backbone, and lays eggs, such as a turtle, snake, lizard, or crocodile.

sediment: Matter, such as stones and sand, that is carried onto land or into the water by wind, water, or land movement.

skeleton: The strong frame that supports an animal's body.

superposition: The placement of layers of rock on top of one another. The lower layers are older than the higher layers.

FOR MORE INFORMATION

Books

Hall, Ashley. *Prehistoric Worlds.* New York, NY: DK Publishing, 2024.

Lundgren, Julie K. *Fossils and Dinosaurs.* New York, NY: Crabtree Publishing Company, 2022.

Websites

National Geographic Kids: Fossil Hunting: A Quick Guide
www.natgeokids.com/uk/discover/animals/prehistoric-animals/a-quick-guide-to-fossil-hunting/
Learn how to become a fossil hunter and what discoveries you can make!

Ology: PaleontOLogy: The Big Dig
https://www.amnh.org/explore/ology/paleontology
Play games, do crafts, and learn more about dinosaurs and the work paleontologists do to study them.